SCIENCE

PUBLIC OUTREACH

PROF. R V M CHOKKALINGAM

ISBN 978-1-63745-029-1

The handbook is dedicated to my late parents @ KUPPAM: My father R V Muniswamy Chettiar taught me to be an independent and determined person, while my mother R. Radhabai showered her love and support - but for them I would not have been what I am today.

Contents

Preface

This book is originally motivated by my desire and stemmed from my passion for Science Engagement, as a champion of Public Engagement with Science for over 50 years now. I am excited to say my five decades of lifetime Science Outreach activities have contributions to build an awareness and appreciation of science, to celebrate the excitement of science and scientific discovery, to improve science communication skills, and to encourage more people to study and pursue careers in science. I am always with progressive professionals, that Science Engagement is a two-way street to have a creative dialogue between community members and science experts about shared challenges and new possibilities. I am delighted to notice now that scientists are increasingly urged to engage beyond academia to improve decision –making, public discourse, and lay understanding of science. Many institutions now encourage Science Engagement and consider it an important component of scientific responsibility.

I am anxious to see the field shift in conveying scientific facts and concepts to main streaming science, and its stories, mindsets, and people in the public imagination. In recent years, there has been increasing recognition of the critical contribution that Science Engagement makes in public arena. I personally feel that there is a profound need to change in engagement settings, styles, and strategies across the public science landscape. The Public Engagement with Science approach and process need to include actors from society, research, industry, politics, and education sectors. This really helps develop a critical public that actively engage and participate in the science discourse to the benefit of society. In addition, it leads to create familiarity with and understanding of the environment in which they live and the science related to this. Public Engagement with science is widely perceived to be a source of many benefits to society and aims at increasing scientific literacy amongst the general public.

I am sure that Science Engagement is at the heart of this book. The book gives a brief overview on the basics, the tools, and the vision behind current Science Engagement practices. The book deals with the basic techniques and methods traditionally used in engaging about science to the public, and a sizeable glimpse of common activities and programs. The book provides with a timely contribution to discussions on public engagement and the associated impact of science. While much of the core guidance seems likely to remain valid, it conveys a sense of norms in the Science Engagement field that continues to evolve. It voices for greater openness and transparency by engagement and dialogue on the part of professional scientists with members of the public. This acts as a reference book for scientists, researchers, policy decision-makers, and other stakeholders. I hope this is well-suited as a handbook and presents options for further reading regardless of expertise.

Science Engagement is an overarching term that includes scientific literacy, science temper, and science culture. The imperative of encouraging a more active and critically reflective engagement with science through the portfolio of activities has been for the meaningful participation, and discussion of policy development. I earnestly state that science promoters need to focus on the need to make science attractive, accessible and relevant through media, public outreach, and promotional programs. Science Engagement helps also scientists to gain from seeing the world outside their lab instead of down their microscope. Scientists should trust the public with their research, and accurately communicate through public engagement projects. There is a definite need in 21stcentury to massif a number of public science programs, and engagement activities to generally improve public appreciation, awareness, and understanding of science to build a knowledge-intensive economy and a better life for all.

PROF. RVM. CHOKKALINGAM

BE(Mech) CDSE(Lond) Dip(Theo) Dip(Astro)

Acknowledgements

I have a long and distinguished career touching many different aspects of Public Engagement with Science for over five decades. I am over helmed in all humbleness and gratefulness to acknowledge the following former directors: Dr. Shanthamallappa, Director, Technical Education, Government of Karnataka, Bangalore for advice and motivation; Dr. A. Bose, Director Visvesvaraya Industrial and Technological Museum, Bangalore for real mentorship and invaluable guidance; Dr. S.R.Valluri, Director, National Aerospace Laboratories, Bangalore for enthusiastic support and encouragement; and Dr. Walter Winton, Keeper, Science Museum, London for respectful interaction and useful discussion. I would like to offer my sincere and great appreciation to Editors of Deccan Herald Newspaper for their continuous publication of my popular science articles all over these years. For understanding my long days at the computer by bringing together the best science resources from across the Internet, I would like to thank my family members Mohan, kumaresan, Kumar, and Manu for their help and assistance throughout the arduous process of writing this book. I express my gratitude to Notion Press for help in the creation of this book.

About The Book

The book is the product of my life-time contribution towards Public Engagement with Science for over 50 years. I have been occupied with commonly run programs offering enormous range of activities to increase engagement with science for over five decades. The book is a hands-on guide for those broadly interested in the craft of Science Engagement and Communication. It is a precise introductory text on Science Engagement with reflections upon a broad range of tightly interwoven fabric of issues. The book makes a valiant attempt to practice Science Outreach with insights into the relationship between science and society. It explores the value of Public Engagement with Science in the wider scientific and societal context. The book looks into various rationales for encouraging greater Public Understanding of Science and research. It highlights the issue of Public Awareness of Science using a wide range of interactive hands-on activities. Content in this book may aid different members of scientific community in different ways towards the goal of Public appreciation of Science. While the contents remain mostly relevant today, the general patterns explored in this book mostly remain true. All concepts and ideas of Popularization of Science are lucidly explained to make it an easy read. The book discuses the philosophy of Science Outreach and the importance of the removal of the disconnect between scientists and the general public. This highly topical book will be of interest to academia, practitioners, policy-makers and the lay public. Let the special handbook take you by the hand and lead you.

CHAPTER ONE

Science Outreach

The modern world would not be modern at all without the understanding of the technology enabled by science. Science affects us all every day of the year, from the moment we wake up, all day long, and through the night. Whether we experience the benefits of the antibiotics that treat our sore throat, or the clean water that comes from our faucet, have all been brought to us courtesy of science. Whether we are experiencing the benefits of cell phone, or turn the GPS on in our cars, we are constantly engaging the benefits of high-tech life caused by science. The advances of robotics, artificial intelligence, and biomedical innovations could significantly alter our human capabilities. Science informs public policy and personal decision on energy, conservation, agriculture, health, transportation, communication, defence, economics, and exploration. Science in the 21st century stands a ready tool to help combat modern difficulties.

Science has served the humanity ever since the beginning and will continue to do so. Scientific inventions and discoveries have made many things possible, which were once thought impossible. Science has demonstrated that it is a supreme mechanism to explain the world, to solve problems, and to fulfil human needs. Science generating solutions for everyday life helps us to answer the great mysteries of the universe. Science has contributed to every aspect of our life and has become one of the basic human necessities. It is not always obvious that science shapes our daily lives, but the fact is science impacts countless decisions we make each day. With the better grasp of the benefits of science, we gain a greater appreciation for science. At present time human life has become totally dependent of science. Every human being has the right to use science for beneficial purposes. Science should be open for the whole society, so it may promote awareness, understanding, and engagement with science among citizens.

Public outreach in science can be defined as any scientific communication that engages an audience outside of academia. The common element of all outreach is engaging the general public in the world of science. Outreach is the nexus between science education, science communication, and science policy. Potential target audience for outreach program include students, teachers, citizens, professionals, resource managers, and policy makers. It is essential to consider and understand how non-scientists assimilate scientific information into their mental frameworks, their beliefs, and their behaviours, Science outreach can have positive effects both on public engagement with science as well as science community themselves. Public engagement, communication, and outreach are areas of activity that can foster curiosity about, and appreciation for science concepts, processes, and applications. The activities aim at increasing the accessibility of scientific information to lay public.

Scientists have a responsibility to communicate with taxpayers about their work and engage with a broader community to share exciting and valuable aspects of scientific process. The one-way communication from scientists to the public must be simplified, sensationalized, and strategically constructed to avoid misinterpretation. Outreach can also include a range of approaches and activities in which information flows from scientists to the public. Current practices in science outreach remain largely based on communication models at many levels. It is by connecting science to the personal domains of everyday life that decreases the metaphorical distance if conveyed through appropriate analogies. The development and preparation for science outreach activities often takes up an enormous creative energy and time. Public outreach is an umbrella term for a variety of activities to promote public awareness of science. The key ingredient for science communication is passion and enthusiasm about science.

There are many ways to bring science to school kids, and the key is to develop projects that take into account the social and economic context of the schools involved. Hands-on and minds-on science experiments for kids, table-top activities, film screening and open days are very good starting points. Science projects and activities help the children think outside their daily reality and hopefully inspire them to aim high. The science outreach called the science fair turns out to be one of the most impactful experiences for the participating public. Other outreach activities include public lectures, field trips, community- based workshops. Science event, science festival, and science talk help nurture a community of practice around science outreach. Museum, zoo, and aquarium have been promoting informal science education for decades. Science toy, science game, and science club are the other means to provide science learning through entertainment.

The Expression Science Outreach indicates increase awareness of the importance and value of science to society. Science outreach activities are greatly needed at this time in our society. They can generate much-needed excitement and interest in science with students and the public. In order to gain the general public to support the pursuit of science, we need to effectively convey its nature and benefits. There seems to be a significant drop in the time, attention, and resources devoted to science education. There is often common attitude among students that science is too hard, too boring, and not worth the effort. Science outreach can help reverse negative attitudes and spark interest and enthusiasm towards science. Science outreach programs are critical to engaging the next generation and exciting young minds to pursue careers in science. There are diverse and great ways to get involved in science outreach and engage the public to enrich them about issues that are relevant to society.

Science outreach come in different flavours. Citizen science enables people from all walks of life to advance scientific knowledge with projects that match their interests. Science nature walk assists students take a trip to the nature to find out examples of living and nonliving things. Family science night is an evening of hands-on science activities for students and their families to do together. Science café is an informal community event giving the public a chance to have discussion with local scientists. Open lab day brings target audience to come and see what is being done, ask questions and may be even participate. Science at con is often a place where the audience really get information on hard science and evolves in terms of content. Such programs feed the public curiosity that makes them cultivate a passion for science learning. The time spent on outreach helps the public develop and enhance critical thinking. Science holds key to the future and provides numerous career opportunities.

Science outreach helps the learner to gain a deeper understanding of the concept through effective communication and to foster a love for science. Summer science camp facilitates teachers to get first- hand knowledge on how science is done. Soapbox science is a program where scientist uses a novel public platform to amaze public with latest discoveries. Science theatre presents an interactive stage show diversely for all ages with multiple exciting and engaging demonstrations. Science Art is a platform for meeting scientists and artistic folks to create scientifically themed art. State-of-the-art science mobile lab opens its doors to families of public as part of scheme. We need to have more immersive experiences in our science outreach in order to improve the quality of science education and strengthen public science literacy. Science outreach strategy needs to be bidirectional reciprocity with communities rather than one-way. Our educational institutions have to articulate a vision for integration with society.

It is widely acknowledged that connecting science with the public is a must. Most science remains publicly financed and, therefore, demands public support. Science never loses its daily relevance with more and more innovations and discoveries surface at an accelerating rate. The mission and motivation of science outreach helps people understand science more effectively. It also helps them know the role that science has got to play in society and in our lives. Public outreach and engagement activities are critical to find common ground on scientific issues concerning the society we live in. Science outreach activities seek to engage and allow children, teachers, and parents to experience science in a fun, hands-on, and exciting ways to stimulate their interests. Scientific organizations, research institutes, and universities have the built in responsibility to get involved and provide funding for science outreach and informal education to integrate research, teaching, and service.

There has been in the recent decade numerous calls issued from various scientific societies for increased dialogue between scientists and the public. Across disciplines, more scientists agree that interacting with the public is a worthwhile endeavour. But participation in science outreach remains fairly low among scientists. More scientists feel that they are unprepared to effectively interact with the public. Common misconceptions probably undermine the value of science communication with the public at all scales. Current practices in science outreach remain largely based on ineffective models of communication. Scientists need to practice outreach with more scientific approach to make it effective. Communicating responsive science to the public should be seen as important as the production of scientific paper. Effective science outreach affords a shared understanding of the facts between scientists and the public. It builds consensus on controversies and foster mutual trust.

Science outreach is a vital connection between scientists and the general public. Raising awareness of how science serves the public is becoming increasingly important. Though scientists have realized this awareness is important, the path forward is not entirely clear to them. Science outreach faces the challenge to prove the value and utility of science, especially to policy makers. The effective communication requires scientists to identify key results of their research, while contextualizing their findings. Public inputs can enrich by offering new perspectives and stimulating inquiries. The public trust in science is enhanced when scientists extend trust to the public and empower the public to assess the data. However, science outreach is now seen as a responsibility that institutions have towards the public. Recently more institutions have become involved in attempt to improve public science education. Several universities run outreach programs with their local schools.

Social media is here to stay, and scientists need to use it appropriately to promote science outreach. Social media has deeply transformed the way scientists communicate ideas and information. It is an important skill to provide science information to the public in a fast, succinct, and accurate manner. Being concise and accessible is the way to make science understood by all. Popular platforms such as Facebook, Instagram, and Twitter can connect researchers with the global public. Professional social media platforms such as ResearchGate, GoogleScholar, and Linkedin allow scientists to network, collaborate, and share their work globally. Proper hashtags, schematics, and analogies can be extremely helpful to explain science to the public. Other effective ways to bring science to the public include short YouTube videos, science sketches, outreach-oriented Twitter or Instagram accounts. Science Blogs is a great example of a repository of science news written for the public.

Television offers an unparalleled platform for visualizing science for the public. Scientific explanations could be supplemented with illustrations and animations. Popular programs include film of natural phenomena, show the insides of laboratories, and even project images directly from microscopes and telescopes. The majority of science television appears in special programs, and special reports. Science broadcasts also now routinely offer information in print and on the World Wide Web, for both teachers and regular viewers to supplement their program content. Prominent scientists have ventured into the mass media to gain public visibility through giving interviews and hosting shows. Books and magazines have continued to be important sources for public about science. The public can now learn continuously from a wide range of print, broadcast, and Internet sources. Science writers are special breed who translate a wide range of science into popular form of public outreach.

Individual scientists still continue to reach the public through more traditional means. The scientists still perceive significant barriers to outreach in an individual level. The computer, the Internet, and the web have made possible new channels for nearly ubiquitous communication. The Internet now makes it possible to obtain information, products, and services both professional and amateur. The web has become the primary source by which the public now finds information about science. One of the outcomes is the dis-institutionalization of education. Podcasts on science topics are available directly from the originator to share research news directly with the public. In addition, the public can now participate with scientists in cutting-edge research. In the future we have to create a world of learning that puts learner at the centre, leveraging technologies and utilizing human capital in new ways. We can forecast this world of learning will be so customized and resilient to engage in immersive experiences.

The Internet is a matrix of networks and computer systems linked together around the world. The Internet is the source of spreading science information quickly to a large audience and of going beyond the limitations of time and space. The World-Wide Web as it became to be known and often called just the 'Web' began in response to a scientific need. Getting science information on-line is easy and convenient. Access to the Web has opened up many aspects of scientific research previously hidden from the public. The Internet is a new approach towards knowledge creation and dissemination called Open Science for a more transparent, accessible, and collaborative scientific culture. Online users are likely to use the Internet almost like an encyclopedia- looking for meaning of specific scientific terms or looking for an answer to a specific question about science. The Internet plays a great role in science communication and has become an effective scientific communication media for scientists worldwide.

Science remains an exclusive discipline as a privilege of a few who understand the language of science, but it is not accessible to all. Science needs to become more inclusive being an inviting discipline for the public. Science has to build a bridge of trust and show gratitude for the support the public have provided to date. The 'how' trumps the 'why' in science and there is insufficient systematic reflection on what all this activity has achieved. Scientific research is funded largely by government and therefore, indirectly by taxpayers. Scientific research can change our lives for the better, but it also prevents risks- either through deliberate misuse or accident. Dual-use research describes technologies that can have both military and civilian uses should be controlled to prevent its misuse by ethical concepts. Science of the future will need to be integrative and holistic, emergent, and evolutionary and profoundly informed by the knowledge of their interdependencies.

CHAPTER TWO

Science Museum

Science Museum or Science Centre is a welcoming and inclusive place with all its hands-on exhibits featuring scientific principles just waiting to be discovered. Science Museums around the world increasingly refer to themselves as discovery centres, making science accessible through their interactive science exhibits. The mission of Science Museums today is to stimulate interest, and creativity by learning through minds-on exhibits for children, family, and the public. Such exhibits delight the visitors with their simplicity in illustrating scientific phenomena. At their best, science exhibits encourage public to experiment and experience real world of science, tickling the curiosity and exposing different phenomena to gain deeper insight. They form an important bridge between formal science education and the community at large. Science Museums create and sustain a culture of learning, which fosters the process of personal inquiry through experimentation, and the sharing of values about the world.

Science exhibits are more than just devices with an on-off switch. They are exciting accents with relevant sounds and visual aids. They do not just explain the physical, chemical, and biological attributes of the world around us, but are live demonstrations of various processes in our world. Hands-on exhibits are the keystone of the museum and education experiences are its main purpose. The reality of science learning from Science Museums is much subtle and complex. The sense of participating in the experiment opens the minds of many. Walk-through interactive displays of physical sciences catch the attention faster than theoretical stuff. The whole point of any science exhibit is to provide the opportunity to investigate and validate the personal theories of young minds directly. Younger generation likes science exhibits where they could press a button, pull a lever, twist a knob, or watch something go pow, bang and whizz. Science Museums or Science Centres are above all for the public.

Science Museum presents kinetic experiences and provides direct multi-modal experiences with multi-faceted portrayals of science using authentic objects and phenomena. Learning should be multisensory, and science exhibits appeal to all the senses equally. Science exhibits are visually exciting, produce sounds, and provide personal experience. To present phenomena that may be hard to visualize for standard interactive science exhibits, they use interactive video techniques, time-lapse films, computer generated graphics, and interactive computer simulations with visual modelling outputs. It is not surprising fact that these exhibits play a critical role in complementing what students learn in school. They supplement the human nature to desire simple explanations for complex reality. Science learning is a dialogue between the individual visitor and the hands-on exhibits through time in Science Museums. Rich, authentic experiences are fundamental to the educational enterprise.

A trip to the Science Museum will provide a rich context about the science and inspire visitor's curiosity. See-for-yourself science exhibits in Science Museums around the world address the desire of a variety of visitors in their pursuit of science. Chosen with extreme care and thought, each exhibit carries succinct labels that are easy to understand, and entertaining; they are self-instructive with graphic aids. Normally the description in the label explains the principle involved and the usefulness of the invention. The labels avoid technical jargon and neither too brief nor too tedious. Kids can explore a process together with their friends while having fun. Joint exploration can happen between students and teachers too, in which the role of teacher and student can alternate back and forth between participants. A relatively free form of environment allows and even demands that children create their own learning path. Science Museums are situated at nexus between the community, the school and the home.

Science Museums and Discovery Centres, however lend helping hand, deconstructing complex ideas and explaining them through working exhibits. Science exhibits at such places are unique resources for non-formal education helping develop skills and positive attitudes towards science. Many Science Centres offer professional development workshops designed to strengthen science teacher expertise and integrate minds-on science activities into the classroom. They often collaborate to develop public outreach and hold programs at the museum or at outside. They organize mobile exhibitions on wheels which are especially helpful to rural masses. They have big parks with outdoor exhibits which are much enjoyable for the general public and children. Science park with their robust displays and exhibits help stimulate and intrigue a wider audience. A visit to the Star theatre or planetarium can be a defining moment- a dome into which a simulacrum of the night sky is projected.

Science Museums have the capacity to create meaningful, socially mediated, and self-directed learning experience. Integrating learning with enjoyable socially mediated experiences enables Science Centres to reach audiences who may not be ready to pursue online. They are places to discover, explore, and test ideas about the natural world. Science Museums have relevance to all sectors of the population and have become important meeting places for science and society. They demystify science, convey its beauty, show its necessity, and make it accessible to the general public. Science Museums are not only places for learning science, but also places of amusement with family, friends, and relatives. Today, there are more than 3000 Science Centres in the world visited by more than 300 million visitors each year. The vast majority of these visitors create durable memories of some aspect of their experience. Science Museums will continue to fascinate millions of minds across the world.

CHAPTER THREE

Science Fair

A science fair is a competitive event where students present their science projects with models that they have created. It is an educational activity for students involving experiments or construction of models in one of the science disciplines. Science Fair is an opportunity for students where their scientific efforts are displayed. The distinguishing characteristic of a Science Fair is that Science Project entries employ the scientific method to test a hypothesis. Science Fair is a great way for students to become more knowledgeable about how the world around them works. Through Science Fair Project students gain a first- hand appreciation of the work of scientists and the value of their discoveries. Most projects also involve a good deal of mathematics and students get an opportunity to advance their presentation. A Science Fair Project is one of the best learning experiences a student can undertake and requires sufficient time to plan.

A Science Fair Project is very straight forward in a sense that the student chooses a scientific question, which he or she would like to answer. A Science Fair Project is an experiment, a demonstration, a research effort, a collection of scientific items. It represents efforts of a student's investigation in some area of interest. The student should explore various sources of information in order to gain a full understanding of the topic. The students' presentations consist of their projects or models along with written reports and display board. Students who have demonstrated their thoroughness in their study and effort are awarded. Many students are far ahead of their teachers in understanding what it is. Most countries have Regional Science Fairs in which interested students can freely participate and winners from here are sent to National fairs. It is an excellent example of active learning or inquiry or hands-on or minds-on learning. Learning about science is at the heart of a Science Fair Project.

The students' presentations consist of their written reports, a display board, and interviews of the students by Science Fair judges who ultimately decide about awards. The display board must prominently state the title or purpose of the project, and the abstract with charts, graphs, diagrams and photographs. All these are to be put together in such a way that the audience can easily understand about the project. The poster should be visually attractive that summarizes the experiment and the result. Boys tend to choose to work in physical sciences and girls in the biological and social sciences. Traditionally the most common type of Science Fair Project is Experiment with significant positive findings and result from the project. Innovation is the next popular type of project consisting of devices, models, and techniques. Study is the other type of project involving the collection and analysis of data from other sources to reveal evidence of a fact, situation, or pattern of scientific interest.

Science Fair is hailed as ode to the teenage science geeks on who our future depends. Science Fair is a long-held tradition that promotes longitudinal interest and self-efficacy in science. Science Fairs provide students with an opportunity to engage in exciting discoveries, learn, and receive potential awards and recognition. The participating students share their interests and connect with like-minded individuals. These young scientists present a vast range of topics from computer projects to hazardous chemicals to reusable energy to autonomous vehicles. Some Science Fair topic areas include biology, chemistry, earth sciences, electronics, astronomy, physics and engineering. The students become ready to make oral presentation of the project after positive and confident work to the judges. Science Fair Projects help develop scientific attitudes and skills, including critical thinking, problem solving, and an appreciation of scientific method.

The motivations of Science Fair participation and the effects are manifold. Many students are sincerely interested in the results of their projects. The positive aspects of Science Fair are increased science content, knowledge, and improved verbal and written communication. The opportunity provided for students to choose a topic and design investigations are the real strengths of a Science Fair. The students could speak in great detail about the processes they used to develop and execute their projects. Their discourse shows that they have a general understanding of a scientific process. Some students indicate that they enjoyed doing the background research and/on conducting the experiment. The other aspect of their Science Fair experience is improving their presentation skills. Some find their participation to be affirming and aspiring in their aspirations for science careers. Students feel that their hard-work in Science Fair is worthwhile.

Science Fairs are so useful and helpful for learning all kinds of life skills. Students show a lot of concentration in doing their science projects. While books provide theoretical knowledge, projects and experiments teach beyond that. Students go above and beyond what is typically required of them in science projects. The experience of devising an experiment, evaluating results, and coming up with a conclusion is new to students. It teaches interpersonal skills like how to present research or how to advocate for your own ideas in a research setting. Presenting and breaking down complex information is a critically important skill for students to learn. Students also discover how to successfully present their work in scientific sessions, and develop skills of being interviewed. The mere purpose of Science Fairs is to instil scientific attitude in the young generation to make them realize the interdependence of science and society.

CHAPTER FOUR

Science Experiment

Experiential Science Learning is a system of philosophy that emphasizes learning from direct first-person experience. It is the most effective way of learn one of the critical pathways toward student learning through exploration, activity, and experimentation. It is all about personal close encounters with the content, process, and emotion of science. It is a holistic perspective that includes the self construction of knowledge. It is one of the critical pathways toward student science literacy. It is inductive, learner-centred and activity-oriented. It is an activity-based learning pedagogy. The emotions, attitudes, and beliefs derived from Experiential Science Education combine to form a learner's science identity. It is the way to break out of the received conditioned teaching practices, which constrain a student's development in schools. Hands-on discovery activities can be done in a variety of ways using everyday odds and ends as the raw material and make experiments more accessible to young children.

The new mindset of low cost teaching aids facilitates innovative teaching paradigms that replace the old fashioned lecture format with hands-on learning. Science experiments are fun and can teach a lot about the world around us. With a little bit of time, and some curiosity as well as imagination, simple physics concepts can be demonstrated by using a few common household items, Simple ideas using readily available materials promote active process of learning science. The simplicity of the effects attracts the kids attention. There is a need for growing use of low-cost teaching aids as settings for pre-service and in-service development for teachers. From cook-book experiments, we have to move towards science enquiry in action. Experience and observation are key to the scientific inquiry process. The students who learn in an interactive environment have a better understanding of science. Such students are better able to solve problems and have higher level of conceptualization.

Majority of rural schools depend on inexpensive teaching aids developed with relevant materials related to the learner's environment. They will also become one of the possibilities to increase the laboratory experience with less cost and sweat. Also this teaching strategy links activity, field, laboratory, library and classroom experiences with real life situations and applications. The learning by doing can be a solo activity or a collaborative effort. The ability to actually make things work is essential for true learning of science. The purpose of low-cost teaching aids is to learn and create an opportunity for valuable and memorable personal learning. The learners are encouraged and helped to learn and develop in their own ways, using methods which they find most comfortable and enjoyable. Low-cost teaching aids not only help students understand science, but also give them a sense of achievement as they perform experiments for themselves and derive their own understanding.s

Do-it-yourself experiments prompt students to interact with them, ask questions and reinforce their own learning. They provide a positive emotional platform for future learning. The basic activities involved in inexpensive teaching aids facilitate children to test ideas, perform experiments, and make interesting discoveries. Both teachers and parents will find them in rewarding ways to provide quality learning experience for children. The functional design of a teaching aid is most important for learning. A constant and intense experimentation using all methods and techniques at one's disposal is very much essential. They are to be created in such a way, where children feel comfortable to explore and learn science. A self made teaching aid out of known materials, when experimented, will help students in quicker and easier understanding of science through first-hand experience. It can be used by classes as props for learning. They need to relate the experiments to everyday things.

The present teaching methods is oriented virtually and exclusively to meet external needs, not people's individual needs and potential. Low-cost teaching aids facilitate active experimentation, concrete experience, reflective observation and abstract conceptualization. They are flexible with open possibilities, and offer integral growth, and discovery. They develop knowledge, skills, and emotions via experience. These learning activities are a means to an end, not an end in itself. For junior schoolchildren, engagement activities in physics can be easily made out of cardboard. Simply physics experiments to kids like soundgun, kazoo tube, anemometer, rainbow spinner, swimming fish, hovercraft, automat, periscope, pantograph, sundial, helicopter, pinhole camera etc. are easy to make, comfortable to handle, and therefore retain excitement. Kids create their own personalized learning paths toward science understanding. Much money can be saved and sophisticated appliances are avoided.

The low cost teaching aids must be very simple and easy to handle. Improvisation and experimentation play a major role in the success of science teaching. Improvisation makes science doing rather than talking. It provides first-hand experience in a variety of ways and cultivates research mindedness in children. It promotes interaction between teachers and students. Improvisation generally adds an interest and involvement in the lesson and encourage cooperative attitude among children. Besides, it helps the students to make use of their leisure time. It attracts attention and promotes sharp thinking, and develops scientific attitude in children. It accelerates the rate of learning and increases the span of retention. Learning experiences appending to the senses are far more effective than abstract learning experience. The science teacher with a certain amount of skill and enthusiasm can replace many pieces of apparatus by an adequate if unconventional, improvised substitutes.

CHAPTER FIVE

Science Toy

Science toys motivate student learning and give teachers confidence to teach science. Science toys are designed to be fun and playful, and are highly motivating for both children and adults. Almost everyone likes to play with toys and such a desire continues throughout an individual's life. Playing with scientific toys, children are exposed to scientific concepts and develop more scientific skills. The use of science toys as naturally engaging resources can provide motivational and experiential links between science concepts and everyday experience both for teachers and students. Science toys have increased acceptance due to their wide visibility and availability, and their own motivational effect. Psychologists inform us that playing with toys is not filling in of an empty period, or just a relaxation or leisure activity, but it is an important learning experience. There is no denying that science toys play a key role in developing kid's interest in science and mathematics.

In a world of Ipads, smart phones, and computers, science toys are becoming more and more important every day. Science toys can be classics or high-tech, cheap or high priced, but they need to engage in spontaneous exploration and discovery. It has been known that learning occurs rapidly in natural way, where children are involved in interactive, and hands-on science toys. Science toys engage a child's senses, spark imaginations, and encourage them to interact with others. Not only will playing science toys help kids learn practical skills, they will also give the confidence to ask questions and accomplish things. They are the way to improve a child's verbal skills but encourage them to talk about what they are doing, yet another skill. Science toys are constructivist that can develop students' inquiry skills. Science toys for kids are stimulating, entertaining, and breed confidence and design skills. They make an explicit connection between being creative and education.

A yo-yo is a classical antiquity toy that was known to Greece around 450 BC. It is a mechanical device for the conversion of energy- a toy flywheel. A yo-yo, in general, consists of a body of rotational symmetry, with a slender axle, which is allowed to roll on a flexible string. Science is a process and science toys help kids explore the 'what ifs' as they encounter while learning. Michael faraday used toys in his 'royal' lectures to illustrate the concepts in 1859-60. He delivered a lecture that included the concepts of gravity and equilibrium. Faraday demonstrated this through a balancing lady toy. The classic balance bird is easy as we can carefully position the bird on its pedestal and then with a light tap, we can make it spin. This spinning continues until friction slows down the balancing bird. Most of us have seen a 'weeble-toy' that wobbles and does not fall down. Even the smallest victory with a scientific toy will encourage children as they will feel as though they have won.

Science toys can be useful for increasing student interest, assisting students to remember concepts, and as the basis of discovery activities to stimulate questioning and experimentation. The use of science oys to teach science reflects an inquiry-based approach to teaching and learning. In learning through toys, students may be able to bring them into share with the class and extend learning and application beyond classroom. In education circles around the world, there is a great cross-fertilization than in may many other spheres of life in learning through science toys. When children play with toys they are actually exploring the world, but on a manageable scale. Children learn what things are and how they work by playing with science toys in direct experience. However, a good toy should be safe and foster exploration. Science toys are innovatively designed to mix and blend education and fun. They help children to improve their fine motor skills as they physically play with hands and fingers.

Science toys are one of the better material resources required in the classroom, with rapidly changing scientific ideas. By introducing science toys in the classroom, children get enjoyable experiences in science learning. So, there is a need to develop science learning settings outdoors and indoors, in which science toys become very handy. In addition to being very educational, the right toy based on science can mesmerise children for hours, as they learn how it works and figure out the nuances of its operations. It is wise to make use of simple science toys as the basis for experiments. The teacher plays a key role in organizing the play environment to help children think about certain ideas. The most interesting and engaging science toys are usually based on the principles of physics. For science teachers, they are great for science classroom demonstrations. By doing experiments that involve children's favourite toys, they can learn a lot about science.

A trip to the shopping mall will take us quickly past racks of science toys, each proclaiming a variety of educational benefits that be gained. Everything from kaleidoscopes, to colour changing mood rings, to UV glowing beads, to fluorescent rocks are great to engage kids with learning about their world. A large variety of science toys are available through toy stores that are safe, cost effective, and demonstrate an area of science. As a product, science toy is compelling to watch and offers opportunities to play and experiment. Using toys to teach fundamental principles of science can be especially effective since toys build on and extend cognitive as well as creative abilities. For teachers and students alike, science toys can provide motivational and experiential links between science concepts and everyday experiences. Teaching science with toys promotes among teachers greater understanding of concepts and commitment to science teaching.

CHAPTER SIX

Science Communication

Science communication is important to keep science in the public eye. It is the relevant channel for essential dialogue between science and society based on fact. Science communication needs to promote meaningful and comprehensive scientific literacy among public to help them become informed citizens who can effectively engage in public discourse. The main goal of science communication are increasing public appreciation of science and influencing policy preferences. Excellent science communication can capture the imagination and spark meaningful debate and discussion that grants science a stronger presence in our society. High-quality, multifaceted science communication builds an open science culture and trust in science and research. Science communication is a complex task and acquired skill. 21stcentury science communication is part of a culture of open science and requires new habits and methods for reaching different audiences.

Science communication is increasingly recognized as integral to the progress of science by engaging public support and encouraging science in society. It is also urged that public participation is part of the role of an engaged citizenry in a democracy. There is no obvious approach to communicating effectively about science. Science communication must therefore be responsive both to people's needs for scientific information and their ways of understanding, perceiving, and using science to make opinions or decisions. The gap between the scientific community and the public highlights the need for re-examining the institutional frame work of communication. Science communication and societal interaction will take place in the bubble of increasingly divergent audiences within a culture of open scientific work. Responsible science communication approach should be taken to using new forms of communication and interaction to reach out to promote societal dialogue.

Science communication is part of a scientist's everyday life. Scientists must become more skilled at communicating in impactful ways. Scientists have to develop the skills needed to successfully communicate about their research with decision-makers and reporters. When scientists are able to communicate effectively beyond their peers to broader, non-scientist audiences, it builds support for science. Researchers could become better at engaging in public discourse by more fully considering the social and cultural context of their work connecting the dots between a scientific discovery and its implications for society is where the challenge begins. For scientists and researchers who work in scientific organizations, effective communication can be as returning a debt created by public support. Communicating science effectively needs more than facts. The majority of the public still respect scientists and value science, but they often find it challenging to discern who and what is legitimate.

The public is made up of all the people who are not experts in a specific field and can differ greatly in their ages, interests, experiences, and opinions. The orientation to science communication means that the needs, abilities, perspectives, and constraints of the audiences are considered in the approach taken to communicating. It is important to be real when it comes to science communication as, otherwise, people will get disillusioned with all the hype and announcements. By thinking carefully about what it is that needs to be communicated, and why this is being done, it is possible to develop high quality activities that are of benefit to both the audiences and the scientists. It is widely accepted that advances in science are most likely to benefit society at large if the public has opportunities to participate in shaping their development. Science can be and in fact, must be accessible to everyone in an understandable format. It is essential to communicate the valuable role science plays and the scientific process.

Science achieves little if it stays in the laboratory and hence its information should not be contained within an elite club. For the last half century, science communication has primarily been the responsibility of teachers, or outreach coordinators, or trained science writers, and journalists with a penchant for translating often complicated science into compelling storylines or concepts easily understood by non-expert public. The media is a mediator between scientists and the public, and they range from print journalists to television broadcasters to documentary film makers. An understanding of how to use media effectively will help scientists successfully communicate their message. While using blogs and other social media platforms such as Twitter and Facebook, scientists must write for general audiences about the research work and share information. It is mutually beneficial for scientists and the public to establish a two-way dialogue.

There is no substitute for science communication to the public and policy makers. To explain specific scientific details to policy makers, scientists must use simple and succinct language thinking broadly of the bigger picture. Policy makers gain knowledge from scientists through meetings, testimonies, and open presentations. Framing science message is important in terms that are accessible, relatable, and meaningful for the policy makers. Effective science communication must clearly and succinctly describe the context of scientific work, its importance, and how the results are useful to the society in general. In order to influence the decisions and policy making, scientists need to learn the art of storytelling to capture the key messages from scientific research and make the data compelling. Engaging in a narrative with broader strokes is more effective to get policy makers hooked to learn more. Understanding science by the policy makers includes appreciation of the nature of its scientific process.

CHAPTER SEVEN

Science Festival

A Science Festival is a public celebration of science and generally involves many different events in many different venues. Science Festivals are, by their nature, rich and diverse experiences, and such have the capacity to meet a variety of learning objectives. Public science events date back to the days of Greeks, when the likes of Plato and Aristotle would speak in public about their theories of science and philosophy. Science Festival is a very fun and enjoyable event for visiting public and a place for them to do science and experience a diverse range of science activities. Science Festival remains committed to deliver serious educational and thought-provoking content in ways that are engaging, interactive, and entertaining. There are now growing number of Science Festivals with people in unique, innovative, and often surprising ways. Science Festival is a way to communicate science between scientists and the public, which is why many Science Festivals have been held around the world.

The World Science Festival, New York, explores and celebrates the entanglement of science and art through a curated program of thought-provoking conversations, inspiring theatrical and cinematic experiences, interactive workshops, and engaging demonstrations holding annually for a decade. Thanks to the increased emphasis in academia on public engagement, it is now expected that learning about science is an open, democratic process- something shaped by professionals, but led by the public. Organizing a Science Festival is a labour of love, fuelled by the passion and enthusiasm. It is a celebration of all things science and its role in shaping our lives and sculpting our future. They have power to engage, excite, and inspire a whole host of audiences from families, to adults to community groups. Science is manifested in an accessible and entertaining way and the benefit of Science Festival is the two-way flow between scientists and the general public.

The World Science Foundation, New York, has the mission to establish and sustain a general public informed by the content of science, inspired by its wonder, convinced of its value, and prepared to engage with its implications for the future. Recent years have witnessed a dramatic global growth in the development of large-scale science events as Science Festival, or Science Carnival, or Science Expo. Science Festivals have been reworking in unconventional and sometimes surprising ways of science outreach. They may include large public expositions consisting of exhibitions, lectures, workshops, discussions, and debates, and both performing and visual arts. Hands-on experimentation and finding things out together can be a great way of building the science confidence of both parents and children. The experience inspires future citizens to participate actively in the decision-making process that affects science and society itself.

Science Festivals are heterogeneous in nature, varying in length and ranging from small local events to large, multi-site affairs. However, they do have certain features in common- all are an intense, transient science experience, and most offer a variety of learning opportunities and delivery styles within their events. Sometimes there may be events linking science to the arts and history, such as plays, dramatized readings, and musical reproductions. Exciting line-up of interactive events and activities that take people on a journey to experience, engage, and explore the power of science. They give people opportunity to meet scientists and find out about valuable work they do to improve our lives, health, society, and the environment. They offer a chance for people to do try out new things via workshops, and experiments. One of the strongest points is that multiple learning opportunities are offered during a visit with literal exploration in the sense of free roaming.

Science Festivals across the world are demonstrating creative energy in finding out new ways to link scientists with audience. They combine science activities, shows, performances, and immersive science experiments. They also offer science shows, musical and theatrical performances, panel discussions, art installations and digital platforms. Science Festivals focus on talks from some of the most well-known scientists with incredible experiments performed in front of the audience. The more successful Science Festivals have shifted away from preaching to the converted towards engaging unsuspecting audiences in unfamiliar experiences. They have a positive impact on large number of people, that they are effectively engaging undeserved audiences, and that they are particularly good at connecting audiences directly with professional scientists. For the public awareness and understanding of science, communicating between scientists and the public is crucial.

Science Festivals with a number of events year around provides the public entry into the world of science. Science Festivals aim is to capture the excitement so that many individuals come away from the Science Festival as citizen scientists. They aim to inspire and engage us in the enterprise of science and discovery. They cultivate curiosity and communicate the power of knowledge and creativity to change our worldview. They promote innovation and cultivate the next generation of global citizens. We need Science Festivals not only in the big metropolitan centres, but also in smaller cities, towns, and rural communities. Some festivals provide science packs containing fun do-it-yourself home activities. People try to share new citizen science projects at Science Festivals and they truly feel like the festival is for them. They engage people from all walks of life in citizen science with lightening strike of inspiration. Science Festivals improve scientific literacy of the public and make them science citizens.

CHAPTER EIGHT

Science Classroom

We are familiar with traditional classroom teaching methods from the years we spent in school. The all for familiar "sage on the stage" approach to teaching is no longer relevant for the modern classroom. The days of classroom, where a teacher desk sits at the front of the classroom and student's desks are neatly aligned in rows are over. The stereotype classroom of today has students sitting in rows, passively listening to a lecture, followed by remote memorization tasks from a dusty, used text book. Information is presented by a teacher using blackboard, overhead projector, chart, video, and so forth. Rather than learning how to think scientifically, students are generally being told about science and asked to remember facts. Scientific experts emphasize the power of technology to influence and enhance academia by providing experience that lead to deep learning. The advancement of technology and innovation are paving the way for an educational space that is interactive, engaging, and fun.

The future of the world is in the classroom today as it is the cornerstone of knowledge, love, understanding, commitment, creativity, and innovation. Retaining school children's attention during a science class can be a challenge, even for the very best teachers. Perhaps this is because there is too much theory and too little hands-on fun for the children to engage in. We have to develop a concept that can serve as guideline for future science education. We have to try to figure out how we can engage and motivate the children by presenting them with the right materials and the ideal design. The children end up being taught science by people who are not really passionate about the subject. The science classroom has to change and continue to evolve with technology. The students have to be able to embark upon a learning path that is customized specifically for them. Children can be turned on to science if and when the teaching follows hands-on research and open experiments.

Today's children are very tech savvy and easily use tablets, laptops, and smart phones to access the Internet, play games, and consume media. Many modern kids are more adept at using various forms of technology than their parents and grandparents. The modern active learning classroom will have flexibility and variety of work and pedagogy, technology, and space to support co-learning, co-creation, and open discussion. They will entail a flexible layout, furniture for utility, technology integration, and light-filled environment. When students are encouraged to take an active interest in what they are learning, they are more likely to retain knowledge they have gained. Regardless of future technology, flexible teaching arrangements allow for easy transitions as students may sit in circle or in pods. They will have teaching spaces to be able to utilize new technologies and flexible enough to accommodate different teaching styles. Adaptable learning spaces allow for the quick and easy configuration

Teachers often form a central point in the jungle of information the students will be paving their way through. Next generation learning is shifting teacher and student roles, that get students in the driving seat. Teachers need to realize that their role in the classroom of the future is morphing: teach from the back and keep an eye on student's screen. Tech savvy teachers must be flexible, communicate digitally, embrace change and use a variety of apps. Teachers will help students to think critically and learn by doing, acting as resource as their pupils discover and master new concepts. Teachers will have to act more as facilitators of learning than lectures. Teachers have to choose and use adaptive learning software in some capacity to allow students the freedom to learn at their own pace. Teachers around the world do seem increasingly confident about the better quality and usefulness of digital tools in classroom conducive to 21st century learning by personalization, and collaboration.

The science classroom has to move away from the old 'chalk and talk' model, or 'one size fits all' approach to more blended learning styles. An increasing number of classrooms have access to interactive whiteboard. It is a rolling touch screen whiteboard that can display websites, images, and videos, which forms an excellent tool for engaging entire classroom in an interactive lesson. The devices must be integrated in a manner that is additive to the learning experience, rather than distractions. They must be accessible to all, safe, and secure, while children need to feel confident using them. By allowing students to use technologies and have it at fingertips, they will learn to appreciate the positive effects it can have on their lives. The rise of personal device use in classroom and even virtual learning opportunities will change the relationship between students, teachers, and technology. Using devices of technology in classroom can help turn traditionally dull subjects into interactive ones.

The potential for using Virtual Reality to teach history, geography, and other subjects is incredible, when we really think about it. Massive open Online courses are easy to sign, and any learner who does not want to spend years in a classroom at school can find courses taught by professionals useful. In the classroom of the future, professors will simply be able to sit and project themselves hologram-style across the globe. The science classroom has to be packed with a wide variety of simple science experiments that enable the children to experiment to their heart's content. Classrooms around the world aspire to more shedding walls, gaining mobility, going smart, and getting ambitious with its design. We need to create new pedagogy by turning to create and provide a stimulating learning and growing environment for children bringing them together indoor and outdoor spaces. The future classroom will have the technology at the centre of this metamorphosis to become relevant for the new era.

CHAPTER NINE

Science Craft

Science craft is impressively a craft with a science component. Science craft ideas are a fun way to combine our love of crafting with our love of science. Science craft can be fun, hands-on, creative, and cool. They are exciting to make and budget-friendly. There are tons of science craft for kids with a plenty of educational possibilities. They keep kids entertained and enthused by igniting their curiosity. Science craft makes for interesting activities for kids who are naturally curious explorers. The artistic beauty of science is all around us and science craft inspires kids to think outside of the scientific box. There is a profound interaction that occurs among the arts, crafts, and sciences. Science craft, therefore, develops such skills as observation, visual thinking, and manipulative ability. Science craft adds value to the pursuit of science. It is a simple way to experience hands-on science learning. Science craft or experiment aims to provide experience in search for meaning.

A curious child is inherently a scientist, who wants to see all those concepts and ideas manifest in real life. There is no better way to learn science than a fun science experiment. Science craft incorporates science into playful activities using common household materials and allows kids to see all those concepts and ideas manifest in real life. It helps enhance learning at home pulling together and using household items we already have on hand. Simple and easy to do experiments are an awesome part of science that allow kids to engage in exciting hands-on leaning experiences that they are sure to enjoy. Science experiments for kids are super easy and a great way to enjoy the world of science. Using different readily available materials, kids make experiments to understand interesting science facts. The amazing experiments at disposal are simple, safe, and easy to follow. Fun science experiments are filled with self-discovery and have a potential of sparkling ideas.

Kids are curious and always looking to explore, discover, check-out, and experiment to find out why things do what they do, move like they move, or change as they change. There are numerous kid-friendly simple science experiments that budding scientists will be stuck in their homes over the foreseeable future. Some interesting examples include: Bend water with static electricity, Set up a Sundial outside, Pulley experiment, Make a Volcano explode, Create a Solar oven, Blow Soap bubble, Put egg in a bottle, Develop invisible ink, Walking water, Rainbow in a glass, canister rocket, Whirlpool in a bottle, Balloon hovercraft, Soap powered boat, Potato clock, Soda Geyser, Build a Parachute, Hula –hoop, Groovy lava lamp, Bottled music, Rubber band Guitar, Sink or Float, Magnet Fun, Smoke ring cannon, Home- made lava lamp, etc. These activities are visually stimulating, hands-on, and sensory-rich for discovery and exploration. Similarly there are lots of easy science concepts that we can introduce to kids to very early on!

From paper bags to cardboard boxes, from paper crepe to thin tissue-children pick any paper for their science craft. The types of paper used include cardboard, cardboard tube, computer printer paper, paper plate, paper cup, waxed paper and drawing paper etc. Common science experiments with paper include paper air foil, weather vane, balancing bird, toothpick top, spinning plate, colour spinning disc etc. Using cardboard, kids prepare boomerang, pantograph, uphill cone, levitator, air-powered balloon, index-card helicopter etc. Cardboard Automata are a playful way to explore simple mechanical elements such as cam, lever, and linkage, while making a personalized science craft creation. Cardboard box is used to build pin-hole camera, smoke-box, seismograph etc. With cardboard tube they construct kaleidoscope, rain stick, tube ball twirler, tube kazoo, sound gun, come-back can etc. A marble run machine is a creative ball contraption made from cardboard material.

Paper is one of the simplest, versatile, available and least expensive materials known to humankind. Although most commonly used for writing, packaging, and wrapping, it is also perfect for making paper airplanes. There is a lot cool science in this activity of paper airplanes such as how forces act on a plane so it can fly. The forces that allow a paper airplane to fly are the same ones that apply to real planes- thrust, lift, drag, and gravity. While the plane is flying forward, air moving over and under the wings is providing an upward lift force on the plane. At the same time, air pushing back against the plane is slowing it down creating a drag force. The weight of the paper plane also affects its flight, as gravity pulls it down toward earth. Science buddies can learn basic principles of aerodynamics- a long flight occurs when these four forces are balanced. Most commonly used size for airplane is A4 printer paper and a variety of flying paper planes like dart, glider, stunt, boomerang and acrobatic can be made for flying.

The most important thing when we are folding a paper airplane is to make each side symmetrical so that each side must match. Look along the length of the plane and see if the front-on shape of one wing is a mirror image of the other. By incorporating a dihedral, the wings are arranged in a flat 'V' to maintain lateral stability. Trimming of paper airplane consists of slightly bending or warping the ends of the surfaces. When the ends of the wings are bent slightly so that one is up and the other is down, ailerons are introduced and the aircraft should turn or roll, depending on the degree of warping. When the ends of both wings are up or down, elevators are introduced and the craft should climb or dive. When the end of the fuselage is warped to the right or left, a rudder is introduced, and the craft will turn. The centre of gravity must remain ahead of the centre of lift for a smooth gliding. The mission of paper airplane is to provide a good time for the child- pilot with the amazement of radically flying through the air.

CHAPTER TEN

Science Project

Citizen Science-Project is that in which volunteers and scientists work together to answer real-world questions and gather data. The idea behind these projects is that anyone anywhere can participate in meaningful scientific research. Citizen Science-Projects are collaborations between scientists and interested members of the public. Odds are there is a Citizen Science-Project that coincides with any hobby interest or curiosity that people may have. Citizen Scientists typically are not professional scientists, rather they are curious or concerned people who collaborate with professional scientists in ways that advance scientific research on topics they are about. Typically the public involvement is in data collection, analysis or reporting in diverse fields. Participants can use mobile phone or the Internet to collect and submit observations to see the results. With today's interconnected world, thousands of people from around the globe can remotely contribute to a study that researchers can use.

Science Café facilitates the public and the scientists to meet in another informal setting with the chance of refreshment. Science Café is the causal interchange of informal dialogue with researchers and the community. The participants treat themselves to a coffee or beer, ask questions, voice doubts, and praise or simply listen and let themselves be inspired. Science Café embodies the mission of creating bidirectional dialogue based on participatory action research. It provides an opportunity to learn, defend, and contribute thoughts on the latest in research and the latest ideas of science that impact society. It is the place for talks and discussions on current issues that encourage critical thinking and open debate. Everyone is welcome to Science Café, which provides a good opportunity for exchange of ideas while sipping at a cup of coffee or tea. There is an important evidence supporting the effectiveness of brief, causal dialogue as a way to increase the public self-rated confidence in science topics.

Science Theatre has a mission to get children and adults alike excited about the wonders of science. It is a theatre performance at which scientific themes or historic events are presented. Science Theatre is a great way to combine culture and research and to convey scientific knowledge in an exciting manner. Further, a fascinating story can facilitate remembering scientific facts which are included into the play. It uses a variety of demonstrations in scientific topics to put together both interactive stage presentations and hands-on activities. The theatrical experience of science is also to show its hidden side, the one that never appears in the official discourse. Such performances can either communicate scientific issues in an understandable and entertaining way or artistically stimulate reflection on a meta-level. The plays deal with moral, political, and philosophical aspects of science. Science Theatre offers play in innovative ways of exploring the intersection of science, history, art, and modern life.

Bird Watching or Birding is a form of wildlife observation of birds as recreational activity. Bird watching has become an exciting hobby that anyone can take part in. It is the observation of live birds in their natural habitat, a popular pastime and scientific sport. The birders come from every social, economic, and cultural group in this inexpensive activity. Modern bird watching is made possible largely by the development of optical aids, particularly binoculars. The bird watching observations are very useful to scientists in determining dispersal, habitat, and migration patterns of the various species. Birds are admired for their beauty, their ability to fly, and most importantly for their role in the ecosystem. Some birds might fly many miles everyday searching for food. The first thing that most bird watchers notice about a bird is its colour, size, shape, behaviour, and song. The song of birds can be soothing and inspiring, but birds sing for more than just the beauty of it.

Theme Park can be seen as a large laboratory to investigate the laws of physics and experience them throughout physically. Theme Park rides teach science/physics and provide one of the possible authentic experiences in science. We can explore physics behind the thrills in amusement parks as the forces our body experiences when we are on them-turns, twists, and rapid acceleration. Allowing people to actually experience the forces in action is not only meaningful learning, but also learning with enjoyment. All of the rides are built with the laws of physics in mind, and it is playing with these laws that makes these rides so fun and easy. Bumper cars are a great place to see Newton's Laws of motion. Riding in a carousel shows the centripetal force that keeps travelling in a circular motion. In the freefall ride, we learn about potential and kinetic energies. Roller coasters are the perfect place to see all these laws, forces, and energies at work. There is an art and science behind those creations of rides associated with thrills.

Science Caravan is a specially designed truck, which carries a consignment of science concepts displayed through science exhibits. It is a special on the road science activity program with imaginatively designed science exhibits that will shock, inspire and popularize everyday science. It is ready to meet young audience groups in regional areas with many activities. It travels to interesting locations, and educational facilities, mainly science halls and schools inviting all interested school children to hands-on science experiences. Science Caravan offers full day exploration through interactive science exhibits for children as well as disadvantaged groups in remote areas. It provides playful and attractive information in science with various interactive media with self-experimental approach. Science Caravan is one of the initiatives of science education path to develop a love for science. The activities stimulate young audience to have more interest in science as a true destination for scientific learning. a

CHAPTER ELEVEN

Science Program

Science Circus program is a comedy show designed to provide interactive hands-on experiences to reinforce basic concepts in physics. One of the aims of Science Circus is to arouse student's curiosity and interest in science designed with a view to arouse fun and excitement. The goal of Science Circus is to focus real-life application of scientific principles through activities of the program. It promotes inquiry learning, engaging questions, and scientific exploration. Science Circus is a beautiful movement and skill and is a great lesson in physics. For example, students learn surprising things about gravity through bowling ball juggling, gyroscopic stability through glass bowl spinning, centripetal force with cowboy lariats, centre of balance with a tall unicycle, and inertia with the old table cloth pull. People watch spectacular Science Circus as the presenters hold fire in their hands, and levitate beach balls. Using the theme of Science Circus acts and their corresponding physics, the lessons are real world based.

Science Club is the place for students to learn about all scopes of science. Science Club is an after school, mentor-based science program for students. This program brings mentors and a group of students to conduct fun and engaging scientific investigations. Schools around the world have a great number of Science Clubs as the first extracurricular that comes in mind. Students get to experiment with a variety of media offering more than facts about science and giving them opportunities to develop a deeper understanding of the world. If a student needs an experiment to answer his questions about the world, Science Club is the right place to look for. Science Club programs can be a great way for students to get out of their comfort zone without too much pressure being involved. It is really just a place to relax and have fun with science activities. Mentors feel that the experience gained by students can influence their creer direction. Science

Club can be simple, safe, and playful.

Science Road Show is a travelling science expo packs in an interactive impressive number of exhibits that will captivate young minds. Science Road Show is a fun , exciting, and interactive learning experience. It includes live shows and hands-on exhibits that broaden students knowledge of science. The program usually lasts for 80 minutes and includes topics for example such as kitchen chemistry, sight and illusion, earth sciences etc. It strongly supports exciting and entertaining learning programs with live demonstrations. Sometimes Science Road Show hosts science shows based around themes and the team uses a range of demonstrations to attract the students. Normally they have popular themes such as astronomy, ecology, environment or physics. For example, number of demonstrations include are like change your shape in the flexible mirror, walk on the floor piano, or shoot a ping pong with a bowling ball. Science Road Show encourages young people to see science in a positive light.

Science Camp encourages all campers to play, tinker, and investigate the world around them with plenty of time to explore. It offers students of all ages an opportunity to really explore science in all its hands-on fun. It falls under the umbrella of what is commonly known as informal science learning. Science Camps come in a wide variety of formats to help teach and introduce science concepts. Students get an opportunity to become junior scientists and embark on a series of science adventures. For students who are interested in specific fields, special lectures by professors or professionals are arranged. Science Camp can introduce many different areas of science and give them the confidence and inspiration to embrace science. Campers are introduced to science concepts through activity-based instruction, including laboratory exercises and outdoor explorations. The benefits of Science Club are broadening their enthusiasm to science and developing the process skills needed in science.

Science Field Trips provide alternative educational opportunities for children and experience more hands-on learning. Traditional field trips include identifying species and sharing photos from nature. Exploratory Science Field Trip users are entrenched in the sights and sounds of forests to know more about flora and fauna. An easy discussion after the Science Field Trip allows the students to freely share their thoughts. Visits to local spots of interest can be an educational and enlightening component of a science course. They are undertaken with a particular reason and purpose and also escape to a usual habit. There are many potential benefits such as improving motivation, interactive participation, and fostering curiosity. Taking Science Field Trip helps science teachers to practice good science teaching and gain science teaching resources. Science Field Trip has the environment of informal science learning with features such as voluntary, unstructured, and open-ended learning.

Aero-modelling is the activity involving design, development, and flying small air vehicles. It is a very excellent and interesting way to learn, apply, and understand the principles of flight. Model Aviation is more than just a hobby that involves small-sized flying objects like gliders, helicopters, boomerangs and other types of planes. Model aircraft are flying or non-flying small-sized replicas of existing or imaginary aircraft using a variety of materials, including paper, balsa wood, plastic, metal, synthetic resin, wood, foam and fibreglass. It is the art of designing, building, and flying non-man carrying aircraft. Powered models contain an onboard power plant either electric motor or internal combustion engine as the most commonly used propeller system. Most of the flying models are radio-controlled and actuated control movement is given on the ground. Aero-modelling helps to understand physics, improve hand and eye coordination, and learn to work with plans and use hand tools.

CHAPTER TWELVE

Science Workshop

Fun Science Workshop for kids offers curriculum correlated, hands-on elementary-level experiments on science concepts. The workshop creates an environment that is visually memorable, fun, engaging, and uniquely experiential. It is an ideal way to reinforce ore science concepts with activity guides. Kids enjoy engaging demonstrations, perform simple experiments, and discover how science helps them to understand the world. They cover very basic topics in physics, chemistry, and biology. Only simple, readily available materials are involved to conduct awesome science experiments by kids. The workshop can be adapted to school curriculum to promote an inquiry-based learning environment. It helps develop a child's resourcefulness, particularly their skills at problem solving. Fun Science Workshop promotes and nurtures intellectual curiosity among kids and help them acquire new ways of asking questions. The experience leaves kids with lasting positive impressions.

Science Teachers Workshop focuses on practical, concrete strategies for common teaching tasks, challenges, and opportunities. Teaching in the 21st century is not only a skill, but also a mindset stuff. The workshop provides training to stimulate innovation in teaching and learning approaches wit appropriate pedagogical skills. The workshop is designed to improve conceptual understanding, instil greater scientific confidence, and provides strategies to help students develop more positive attitudes towards science. It draws on research-based best practices and helps teachers consider ways to apply them in their teaching. By participating in hands-on activities or exercises, the teachers understand the processes and skills scientists use. The training program is made to be interactive and case-driven. It provides professional development for science teachers in teaching physical sciences. Science Teacher Workshop is great for exposing teachers to new ideas, but true change in the classroom requires long-time work.

Community Science Workshop is a place for all to tinker, make, and explore their world through science. Community members of all ages engage themselves in a variety of activities in a flexible place packed with gadgets, exhibits, and tools. The core program is a permanent, dedicated physical space, full of interactive hands-on physical exhibits as well as tinkering and making space. There is a whole bunch of programs to disseminate the science with the central workshop space, which is at the heart of it. Given a wide variety of stuff, they will find something that they are excited about, and will take on projects of their own interest. Community members grab materials and tools to make and tinker their own creations as part of practical work. They build things over time and get involved in different projects for different reasons. They find it comfortable with the shared space and events. At the end of workshop, we can notice user-generated content mostly hand-made.

Teaching-Aid Workshop is an experiential workshop dedicated to helping science teachers learn how to make Low-Cost Teaching Aids in physics. The main objective of the training program is to enable science teachers to make inexpensive teaching aids useful for classroom teaching. The science teachers really have a chance to practice the techniques and skills to make simple experiments using hand tools. In order to interest the teachers, the jobs given are made simple and easy. The basic teaching- aid design takes into account the raw materials and technical skills that would be available with rural science teachers. The teachers are guided in the construction of apparatus as per the project drawing of the manual along with prototypes. To a relative extent, the creative work of teacher to make things with their own hands has been stressed with increased ability. The development of Low-cost Teaching Aids would be extremely rewarding as it brings both students and teachers together with the realities of science.

Science Dialogue Workshop is designed to prepare university and industry researchers for successful interaction with youth and community. Science expertise does not always translate to successful engagement in free-choice learning environment. The idea of the workshop is to introduce oneself and one's research using simple terms and easy-to-understand language. It focuses on motivation and relevance to solving mysteries and meeting challenges, encouraging questions and concerns. It provides an opportunity to foster face-to-face engagement with broader audiences. Science Dialogue Workshop attempts to better prepare scientists for effective and rewarding outreach experiences with the general public. The participants learn how to better present their research with new perspectives to a wide range of audiences and gain increased skills and confidence in engaging in conversation about science with them. These skills also translate to networking with other scientists outside their specialities.

Science Policy Workshop exposes researchers to the science-policy-interface. This workshop aims to provide training to researchers to raise their awareness of the need for communicating beyond the research community. Communicating scientific information to policy-makers requires certain skills in translating the scientific information that can be understood by policy-makers or non-scientists. The knowledge transfer from scientists to policy-makers plays a major role in modern science and will be even more important in the future. It gives participants the chance to find out more about the policy process and the methods needed. Science Policy Workshop covers areas such as policy for science (research funding), science for policy (use of scientific evidence in policy), and science diplomacy (role of science in policy issues). The dialogue between scientists and policy-makers is crucial to ensure project finance based on sound scientific knowledge and natural collaboration beyond science.

CHAPTER THIRTEEN

Science Fiction

Science Fiction is just that fiction about science, and science cannot be taken out of Science Fiction. There is an undeniable link between science fact and the ideas that emerge in Science Fiction. Science Fiction expands the theories being worked on now, and explores what may be possible in the future. It is important to note that Science Fiction has a relationship with the principles of science. It frequently builds on scientific developments that have already captured on public imagination. It contemplates possible future, a means of thinking about reality. Science Fiction draws inspiration from current events that are important or interesting, and tell the story in a way that removes of the biases that people may have. In other words, it means creation of an engaging world with enough scientific depth. Science Fiction is not grounded in what we know today, but it is a great source to what can be tomorrow. For scientists, it is an inspiration machine, and offers a kind of exercise for the imagination.

Science Fiction makes us think, wonder, and ask what if and why. It is interesting to note that Science Fiction is often the first foray into new ideas with limitless possibilities. It is always been a way to explore what our future might look like, and Science fiction ideas inspire us to come up with new inventions and make it reality. It can often look wildly different than the world we know, but that is rooted in the world we know. Science Fiction is at its best when it teaches us about the world that we live in and the future. Science Fiction has always been concerned with social issues and gives a message that people can understand. It helps us think ahead, and predicts future technologies, but it creates debate by asking what if? We may not immediately believe it, but future technologies are a lot closer to Science Fiction than people think. Perhaps its strength is how real they make everything feel. We may be correct to expect that any technology represented in Science Fiction will eventually find its way in the real world.

Science Fiction writing is a craft that requires both talent and required skills. The author needs to read widely and voraciously. Great Science Fiction begins with an idea. The writer tries to develop a really good story based on his new idea. The more research the author can do into his selected area, the more ideas he will have. The way what if element operates in Science Fiction is based on the principles of actual science. Science Fiction writers often seek out new scientific and technological developments in order to prognosticate freely the technological and social changes that will shock the people and expand their consciousness. The authors speculate about alternative ways of life made possible by technological change and it is sometimes called Speculative Fiction. They often include a human element, explaining what effect new discoveries, happenings, and scientific developments will have on us in future. The works of Science Fiction writers has to be plausible, reliable, and yes it has to be real.

Science Fiction film is often set in the future, and entertains to chart the dreams of our future. Sometimes it is the seemingly weird ideas that come true by sparkling an imaginative. Science Fiction film is unique genre, which creates much more than ordinary reality by blending technology and fantasy to create a world in the imagination. There must be an element that makes a feeling this is possible by motivating advances in science and technology. The classic elements of a Science Fiction film includes: Time Travel, Teleportation, Telepathy, Telekinesis, Aliens, Space Travel, Parallel Universe, and Fictional World. They emphasize actual, extrapolation or speculative science and he empirical method, interacting in a social context with the unknown. Science Fiction movies are usually scientific, visionary, and imaginative. They usually visualize through fanciful imaginative settings, expert film production design, advanced technology gadgets, scientific developments or by fantastic special effects.

Several Science Fiction films have foreshadowed modern day technologies and so Science Fiction is faster becoming a science too. Several Science Fiction films have very accurately predicted and paved the way for many of the pieces of technology we enjoy today. They offer a unique approach to thinking longer terms about technologies, and examines the possibilities and implications of new technologies. Almost every new technology that becomes practical was previously used in Science Fiction films- Artificial Vision, Transparent Aluminium, Tricorder, Repulsor Beam, Instant Messaging, 3-D printers, iPad, Google Glass Flip phone, Hover-bike etc. Awesome technologies inspired by Science Fiction film include- Flat

Screen TV, Video Call, Lightsaber, Human Teleportation, Time Machine, Faster than Light Travel, Gravitational Shielding, Infinite Data Processing etc. It is indeed our expectations of technology in the real world are fed by our perception of Science Fiction content.

Science Fiction is mythical, magical, or religious, but it is scientific. In Science Fiction films reliable or believable science is part of the plot and science is so much integral to the plot that cannot be removed from the story. Science Fiction film stories are set in the future, or in space, or on a different world, or in a different universe or dimension. It is the details that sell Science Fiction film as it makes the world seem real by ensuring the story constantly rings true. Science Fiction film is an optimistic genre, inspired by the same thing that has inspired the greatest science discoveries throughout the ages. It is widely accessible, invites the exploration of ideas, makes it easier to understand complex ideas, and wants to get its science right. Science Fiction has produced some of cinema's boldest and glorious flights-in every sense. All Science Fiction films have elements of action, drama, adventure, and mystery without blurring the lines, and they play such a key role in shaping public opinion.

CHAPTER FOURTEEN

Science Sparkle

Science is deeply and widely rooted in everything we do today, including common activities of our daily lives. Almost everything that makes ease of our daily lives are the wonder of modern science. Science has eased our lives so much and now it has become almost impossible to live without using them. We encounter science every single day: there is physics in cameras and cell phones; there is chemistry in food and drugs; there is biology in healthcare and medicine; and there is mathematics in recipes and fitness. Truly science has given ears to the deaf, eyes to the blind, and limbs to the crippled. Science responds to needs and interests of the society in which it takes place. Science is important because it influences most aspects of everyday life, including food, energy, medicine, transportation, leisure activities and more. Every citizen will admit the enormous impact of science on the development of a more healthy, happy, prosperous, and knowledgeable society.

Science contributes to ensuring a longer and healthier life, monitors our health, and provides medicine to cure our diseases. There are so many things around us that we take for granted, but make our lifestyles possible through science. While the importance of science in our daily lives may not be obvious, we actually make countless science-based choices each day. A basic knowledge of science can help us understand many of things that we use or that affect us in our daily lives. The effect of science is felt by every person in every sphere of life. Science is not merely a reflection of human progress, but rather an essential contributor. Science has provided the foundation upon which have been built the modern society. It is the right thing to encourage participation in science from as many sectors of the population as possible. It is important to ponder new ways in science, generate new ideas, and share with others, so the concept of science for the benefits of all remain alive for ever.

Science generates solutions for everyday life and helps us to answer the great mysteries of the universe. Science creates new knowledge, improves education, and increases the quality of our lives. Science and innovation drives our pursuit of more equitable and sustainable development. Science is deeply interwoven with society and both science and society are changing. Modern scientific practices have been transformed by increasing knowledge, changing societal concerns, and advances in communication, and technology. With the advancement of scientific knowledge, the question we seek to answer have become more complex and by the side science has become more specialized. The specialization has necessitated more cross-disciplinary collaboration than ever in the past. Science needs to become more multi-disciplinary in order to promote integration between the society and science. A holistic approach demands that science draws on the contributions of humanities and cultural values.

For a sustainable development challenges, government and citizens alike need to understand the language of science and must become scientifically literate. Public understanding and engagement with science, and citizen participation through the popularization of science are essential to equip citizens to make informed personal and professional choices. The possibility for human enhancement stem from new scientific and technological innovations. Science and technology that are directed towards the good of humanity are indeed praiseworthy achievements. At present time, human life has become totally dependent on science as it has contributed to every aspect of our lives and has also become one of the basic human necessities for existence. Science drives much of the world's innovations and the faster the science moves, the faster the world moves. We are influenced by the culture in which we grow up and the society in which we live and hence our science.

Science is a global human endeavour and a collective enterprise of researchers in successive generations. Science is a multi-layered complex system involving a community of scientists engaged in research using scientific methods in order to produce new knowledge. Science is also a process of discovery that allows us to link isolated facts into coherent and comprehensive understanding of the natural world. Science is also the foundation of an innovative culture and can be found at the core of significant political decision. By learning to be more scientific thinkers, we respect the facts, question our beliefs, and practice our knowledge in the real world and so never stop learning new things. Science is the foundation of our culture and society, and so understanding science is crucial to know the impact on our future. The achievement of science is to know new things, while the evolution of science is to know them in new ways. At the core of science's self-modification is technology.

Science allows us to develop technologies, solve practical problems, and make informed decisions. Science often tends to change the whole world wit just inventions and innovations. Science helps us develop critical thinking, reasoning, and decision making skills that will serve for life-time. Scientific knowledge may factor into our everyday decision making. Science helps us to develop technical skills with the understanding of the safety considerations. It provides energy, and makes life more fun, including entertainment, music, and sport. The infrastructure of society as known today is the result of science. It is science that allows us to understand how we learn music, how we move our bodies to dance, and how our eyes see it. Humanity will change more in the next two decades faster than ever before. Science is the main tool to explore the hidden secret of the world for the betterment of human life. Science can also help us connect concepts and explicitly link different disciplines of science.

Science drives much of the world's innovation, and almost every technological and medical innovation in the world has its roots in a scientific paper. Science is in a transition period between a pre-web form of communication to a natively web form of communication. Scientific content will increasingly become native web content created with the full interactivity, and richness of the web in mind. The World Wide Web has exploded to more than 15 million pages that touch almost all aspects of modern life. Around the world researchers are increasingly aware of the value and importance of open science. The Internet has become an omnipresent utility, something we expect to always be available and around us, intertwined in our daily lives. The digital literacy and access are now common to all who teach and communicate their science and to the audiences. The astonishing pace at which social and digital media have permeated every aspect of life is the real gift of science.

The advent of the Internet delivers unheard quantities of information to people around the world. We can access anything we want all at a few clicks of the button. E-mail has led to instant messaging. Digital data is profoundly networked and ephemeral in ways that were simply not possible before. The mobile device has become the primary connection tool to the Internet for most people. Science affects almost everything we do today and it also influences most of our plans for the future. We can now instantly access data and services via the touch of a screen, order shopping, rent cars, plan our journey to work, and book doctor's appointment, all from our hand-held smart phone. Augmented reality and wearable devices will be implemented to monitor and give quick feedback on daily life, especially tied to personal health. The future technologies include Internet of Things, Artificial Intelligence, Virtual Reality, Robot in space and in the workplace, Self-driving car, Hyper-fast Trains etc will change our lives forever.

The notion of the citizen-scientist has increasingly been highlighted in many different contexts, in which the reciprocal partnership and engagement among researchers, citizens, and policy-makers is recognized as the key to the success of multi-stake holders initiatives. Linking science to society, public understanding of science, and the participation of citizens in science are essential to creating a society where people have the necessary knowledge to make professional, personal, and political choices, and to participate in the stimulating world of discovery. Public participation in science initiatives in healthcare and the environment can open up new directions in research. Public engagement, whether it is face-to face or over social media, will encourage people to think actively about science and understand its relevance to them in their daily lives. Increased interaction with the public will effectively inspire new champions for science research and encourage them to lobby for science.

CHAPTER FIFTEEN

Science Engagement

Scientists around the world have made remarkable progress toward understanding the human body, our planet, and the cosmos. Researchers note a lack of time and money for public engagement, besides mentoring, and training in engagement techniques. With a dramatically changing media environment, challenging economic and social climates, there is a shifting relationship between citizens, science, and policy-makers. Being able to communicate effectively with our society helps scientists be transparent about their work, change attitudes and opinions, and justify what they do. It is equally important for scientists and researchers to be aware of public attitudes towards the potential applications of their research. The focus has now been on the best way for scientists to engage with the public or how to influence policy-makers. The more public engagement is practiced, the cleaner becomes the tangle of institutional motivations behind scientific research.

Public Engagement with science facilitates dialogue between scientists and the public about the benefits, limits, and implications. It works to further the public awareness of science and the scientific process, and increases public input into scientific research and policy agendas. It increases the public understanding of science to establish dialogue between policy-makers, the general public, and the scientific community. The public engagement with science convenes dialogue between science and society to draw on relevant information and expertise from multiple perspectives. It takes communications approach in which scientific information can be deposited with the public in order to be literate in science. Community engagement means hearing from a large and diverse audience and collecting meaningful public input to inform decisions with an audience outside academia. We share a normative commitment to the idea of democratic science policy and hence Public engagement can be a part of this.

Science Engagement is by its nature purposeful and at its best supports people to make connections between ideas, their own lives, and the world around them. Science's new social contract with society must now ensure that scientific knowledge is socially robust, and that its production is seen by society to be both transparent and participative. The public input into science-related policymaking is an important facet of Public engagement with science, and typically oriented towards achieving actions and outputs from the interactions. In public dialogue approaches, the goal is to recognize that informed discussions with the public can result in learning by both the public and experts. University-led engagement has the emphasis on trust-building and social learning in collaboration with key stakeholder groups. But this requires a more integrated approach to science communication and engagement across the public.

Science is prevalent in all facets of our lives, and the science-society relationship is complex, and there are many ways to approach. To fully realize the social, economic, and environmental benefits of the significant investment in science, research, and innovation, we must communicate and engage the wider community more fully in science and in understanding of the knowledge economy we aspire. Science engagement can cover a broad range of activities including public engagement around research, teaching-led engagement, schools and outreach activity, collaboration in enterprise or activities undertaken by the university. A growing body of work suggests scientists benefit from engaging with the public. Further, involving a wide range of interested stakeholders can connect seemingly unrelated viewpoints, with potentially far reaching effects. This evolving culture of Science Engagement is an initiative to explore how the public connects to science, and how science connects to the public in the 21t century.

The purpose of Science Engagement can serve include Informing (inspiring and educating and collaborating); Consulting (actively listening to the public views, concerns, and insights); and Collaborating (working in partnership with the public to solve problems together, drawing on each other's expertise. The events and activities include: Hands-on Experiments and Demonstrations, Science Exhibitions and Festivals, Science Film Screening and Theatre Performances, Science Workshops and Training, Science Debates and Discussions, Science Po-up spaces and Games, Science Fun Make-and-Take Activities, and Science Cafes and Citizen science Programs. It also includes apps, websites and discussion forums. A coordinated constituent and effective science communication and engagement policy and implementation will make a meaningful contribution to bridging the gap between science and society. The value of crafting Science Engagement needs to address concerns and motivations of the public.

Science Engagement means a two-way process inviting people to actively participate and engage with ideas and concepts. The two-way communication approach is a common point amongst the most recent and innovative definition of public engagement. Facilitating the involvement of the public can inform and improve public policy and allows government to make decisions that are responsive to the needs, and the will of the public. When people find some proximity to science or finally get how interesting, or fun, or unusual science can be, that is when it is all worth it. It is about creating opportunities and experiences for all people to engage with science on their own terms. Several studies show a general positive correlative between high-quality community engagement with science and positive attitudes towards scientific research. There is nothing more rewarding than to see people use science to make their lives better, and to realize science as an important agent to change their lives.

CHAPTER SIXTEEN

Science Media

The media is an important source of information for the public and can influence public's perception, attitude, and behaviour. Public science is basic scientific research funded by government. There are challenges involved in making science public, or making public science, or making public in science. Most scientists consider visibility in the media is important and responding to journalists a professional duty. Good science reporting should be accurate and should put research conclusions in context. Translating complicated concepts that are jargon-heavy into terms and ideas the public can understand is not always easy. Scientists need to write and share their work with the public in accessible ways, when the magic happens. Scientists can build bridges between research and society, engage the general public, and develop a critical dialogue about the solutions science offers. Some of the most complicated and urgent public policy debates been questions of science.

It is imperative that the public is engaged in science issues which have an impact on their lives, and in their own self-interest to thrive in modern society. There are also other ways to disseminate academic research to the public, including writing research and policy briefs, sharing it on university and research institutions, websites and blogs. Most importantly, always remember to keep everything short and to the point, because many media outlets have a word limit. It is important to find the story and storyline in the research by creatively linking and connecting to the global challenges facing humanity. Metaphors are an essential part of science, from doing basic science to engaging in popular science communication. There are just a few of the media subjects that have significant implications for both public policy and personal decision-making: pandemic spread, global climate change, stem cell research, space exploration, renewable energy technologies, bioterrorism, genetic engineering etc.

The general public gets most of scientific, environmental, and health information from news media. Science news ranges from basic research to applications of science and technology, as well as society responses to scientific developments. In an usual way, important science news would be covered as it happens in the daily news pages. There seems the number of important science and science-policy developments need to be covered increasing. Scientifically illiterate are woefully unprepared to understand basic scientific concepts or applications of science and technology. The role of the news media in conveying the latest information about science and public policy is crucial, providing with frontline coverage of current controversies facing society. The scientific community, particularly those receiving public funding, have an obligation to make communication with the public through the news media a valued part of their job. Science must make the media its ally for implementing widespread change.

The development of science writing as journalism specially mirrors the growth of the scientific research enterprise. Science reporting is often characterized by a fascination with the new developments in science, medicine, and technology. Science writers would cover spot news and write in-depth analytical piece that compete for the front page. Print journalists at major newspapers, magazines, and journals provide more in- the need for compelling visuals. Many important topics are left unreported in favour of soft news, we can use such as consumer health and medical features. There is a greater need for journalists to report both the underlying complexity of science and technology as well as the legal, ethical, and political ramifications. Science writers themselves need to be stronger advocates for science and technology coverage. Science sections have to continue in number and size among smaller newspapers.

In recent years the Internet has provided a new venue for public access to both scientific developments and writing about science as well as opportunities for citizen journalism. Online sites, blogs, and social media tools are becoming the twenty-first century version of the science section and are already replacing or augmenting the print version science sections in newspapers and magazines. Online offers the print and electronic news media opportunities to expand their web coverage of specialized topics in science and technology. In the electronic media, television network and cable news coverage continues to be in a limited way. The Internet does provide helpful, accurate, and engaging coverage of science and technology that is accessible to a wide range of audiences. Surveys indicate that television is still the main source about science and technology in its myriad forms. The online and social media sites tend to appeal to niche audiences, rather than the broader audiences.

Science and media neither really understand each other, and yet they both need each other to progress science and sell newspapers. Scientists and journalists are themselves have to be well-prepared for communicating about science and public policy in a manner that helps the public better understand pressing issues o f science. Science reporters are required to cover the whole gamut of science from the basic research to the public-policy implications of the use and potential misuse of science. Science stories have to lay a growing emphasis on the intersection between science, policy, and politics. The news media must help sort out the potential public and personal choices facing both decision-makers, and individual citizens. The challenge ahead is to boost the coverage of the important issues in science, providing the important insight and context for understanding the stature of scientific research. Science is everywhere in society; a part of each person's everyday life.

CHAPTER SEVENTEEN

Science Everywhere

Science has provided the mankind with remarkable insights into the world we live in. Right from the invention of wheels to building super computer, science has made our lives easier. The wonders of science that have benefited human beings are almost countless. Scientists have made use of science for the benefit of man and society. They have used scientific principles, laws, and facts to invent and develop machines of different kinds. Science has become an exclusive profession, which needs long years of devotion, training, and apprenticeship. Science gives wings to our imagination by its facts and theories. Science is completely based on facts and experiments, and so basic knowledge of science is mandatory for everyone. Science is supreme knowledge, and it should be used for the supreme good of all human beings. Science and research help to satisfy many human needs and improve living standards. We can see the use of science in each and every aspect of our lives.

Science ideas of doing something in a better, or easier way, or of using less of our muscle power has always been a goal of humans. The role of physics and its principles play in our lives and the way man has become aware of these principles throughout history, utilized them to better our lives. Science has shown us how we can do a greater volume of work in shorter time, and with less physical strain. Simple machines were the first ones created, and we still use them today. We all do work in our daily lives, and we all use simple machines without even thinking about it. The most notable of these are known as lever, inclined plane, pulley, wheel and axle, screw, and wedge. Our kitchen is full of simple machines-things such as knives, tongs, scissors, bottle openers, forks, cheese graters, and vegetable peelers. Nail clippers are a simple form of compound lever, while the edge of blades act as wedges. Every home today has all sorts of simple machines without which many tasks would be almost impossible.

The natural world contains an infinite variety of patterns constantly created by simple physical laws. There are physical and chemical reasons behind incredible visual patterns in the living, and non -living world. Patterns are outward manifestation of an ordered structure and are clues as to how things are organized and connected. We see a great diversity of patterns and observe mathematics behind the symmetry of nautilus shell, honey comb, sunflower, peacock feather, snowflake, spider web, zebra stripes, and leopard spots. Natural patterns include symmetries, spirals, meanders, waves, stripes, and cracks. In biology natural selection can cause the development of patterns in living things. The laws of physics show bilateral, or mirror symmetry in living things mainly in animals, as do the leaves of plants, and some flowers. The designs and structures inside fruits, vegetables, and flowers will surprise and fascinate us. Patterns in nature do not stand alone, but as part of a complex web of interrelationships.

Computers are now a fact of life, and they have become heart and backbone of society today. Computers have ultimately altered the way today's society works, communicates, entertains, and educates. Every small action performed in a job eventually goes through some kind of a computer. All businesses use computers to keep track of accounts, money, or make transactions. Computer allows one to easily send and receive emails with just a click of a button. Today, we can attend school completely online, never having to step foot outside of our homes, or attend both online, and on a college campus. The convenience of computers is that we are able to access the computers 24 hours a day, 7 days a week, and 365 days a year. We use computers to our full potential in order to do more tasks, and to do them at a faster pace. Computers also benefit us to do more effective treatments for a healthier and longer life. Computers benefit society with the enhancement of knowledge and skills.

Science has a major role in the larger world of leisure and entertainment. Entertainment is so important in our lives and plays a very big role in living normal and happy life. From early radios and music through to Virtual Reality, science is embedded in all aspects of entertainment. There have been key scientific developments in areas of Audio, Radio, Television, Photography, and Movies. Entertainment science has provided diverse products and services including print media, toys and gaming. Motion pictures, home videos, and television programs, music, sound recording, and videogames are collectively one of the largest and fastest growing economic sectors. Entertainment- education has a strategy for containing educational messages into popular entertainment media with the goal of positive awareness, knowledge, attitude, and behaviour. Entertainment has always been associated with amusement, enjoyment, relaxation, and fun. Society itself is simply interested in science in all its manifestations.

The world has already benefited greatly from space science and technology, especially in terms of communications, positioning services, earth observations, and economic activity of space programs. Space programs have led to the development of various technologies such as GPS, accurate weather prediction, solar cells, or the ultraviolet filters in sunglasses and cameras. Satellite technologies have already revolutionized banking and finance, navigation, and everyday communication, allowing international and long-distance national phone calls, video feeds, streaming media, and satellite TV, and radio to become completely routine. Space-based location services built into mobile phones, and used by applications on mobile phones ranging from maps to dating services have become so intertwined with modern life. Space spin-off technologies have found beneficial uses on earth including cordless power tools, lightning detection, solar panels, heart monitors, advanced robotics, and energy storage.

Author Bio

Prof. RVM. Chokkalingam is a former Lecturer/ Curator/ Scientist and now @ 77 is a local professor and a consultant in Public Engagement in Science. He started his career as Lecturer @ Sri Jayachamarajendra Polytechnic Bangalore (1966-68); worked as Curator @ Visvesvaraya Industrial and Technological Museum, Bangalore (1968-83); and became Scientist @ National Aerospace Laboratories, Bangalore (1983-2003). He is a Science Museum Scholar, specialized in 'Design of Science Exhibit' @ Science Museum, London. His long and distinguished career in Science Engagement and Communication spanned more than five decades. He is one of the pioneers in the development of 'Low-Cost-Teaching Aids in Physics' to help equip rural schools. 'Science- on- Wheels' has been his initial opportunity to design and pilot hands-on science exhibits. He mentored school students and teachers in creating innovative 'Science Fair Projects' and now continues to be a jurist for several Science Fairs. He is a 'Science Writer' and has authored several popular science books and published many articles in newspapers and magazines. He is the recipient of Karnataka State award for 'Science Communication' in 2012. He is a paper airplane enthusiast.